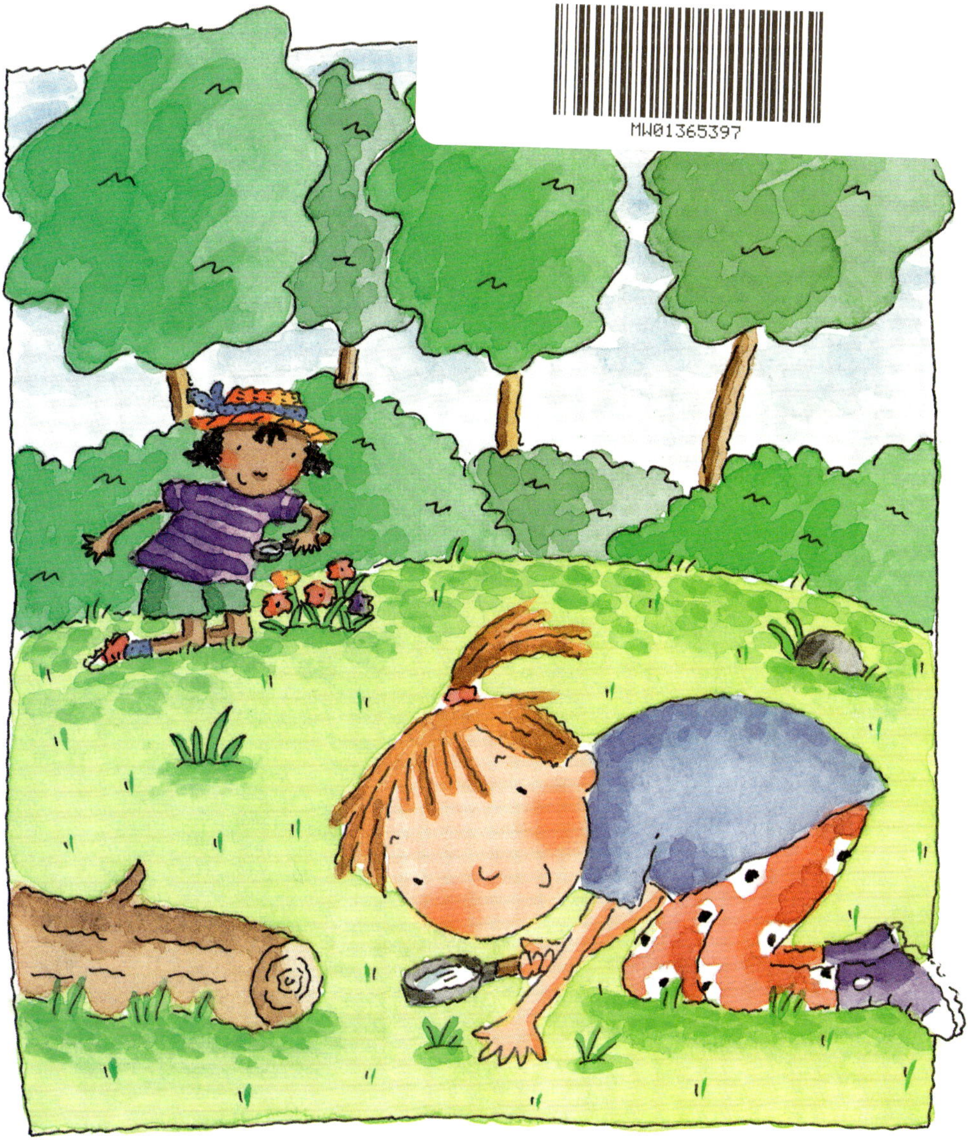

I hunt bugs.

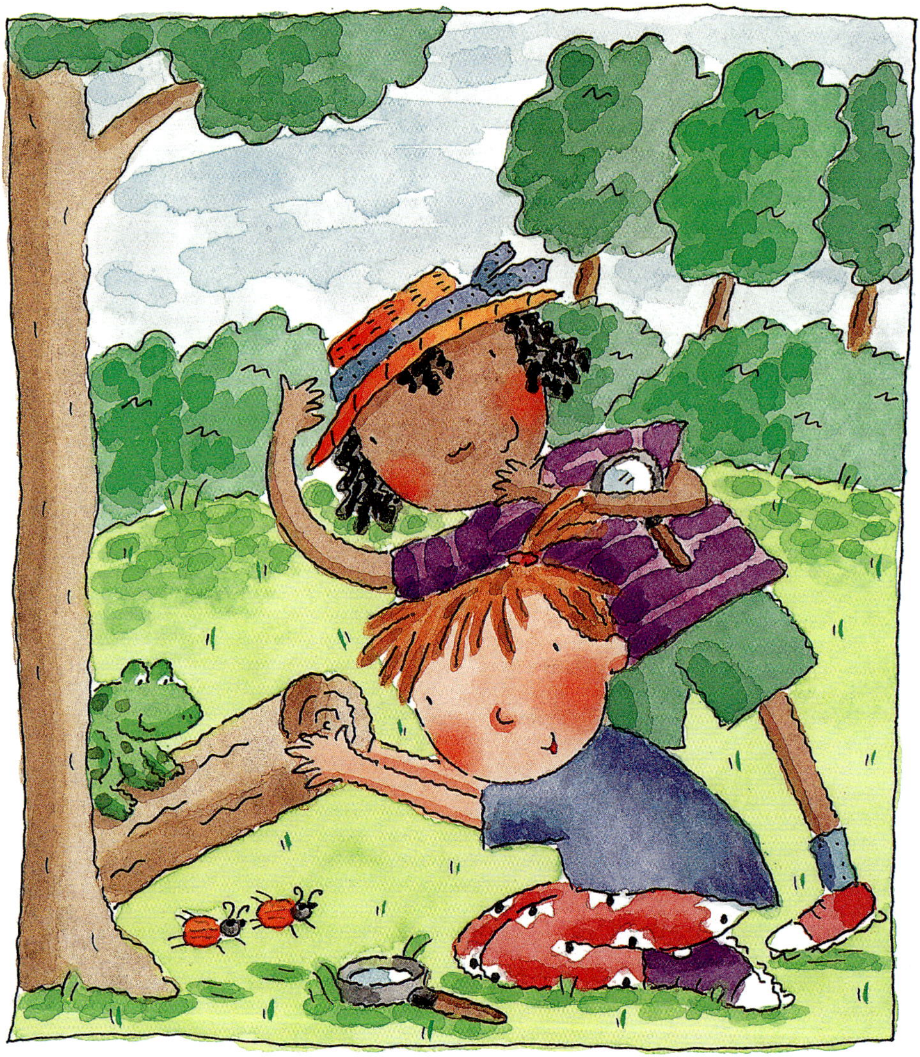

Bugs run, but not from us.

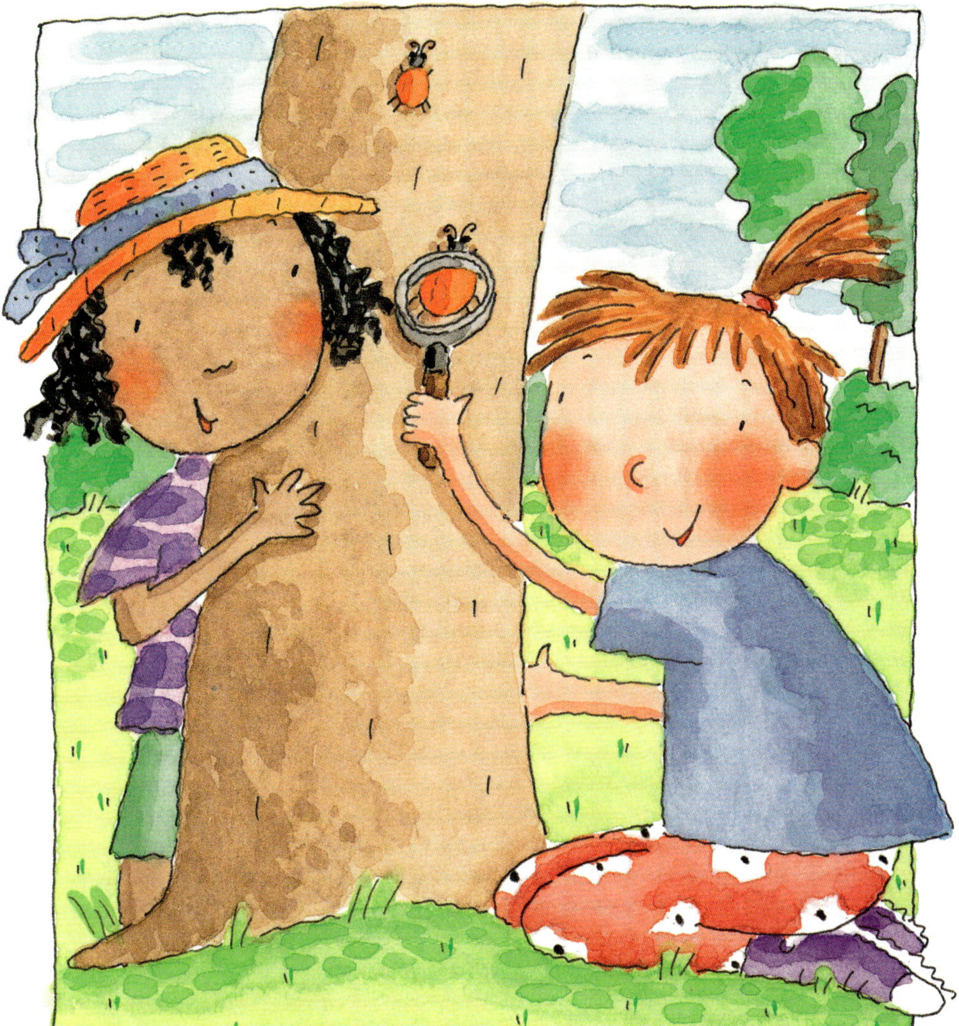

The bugs run up and up!

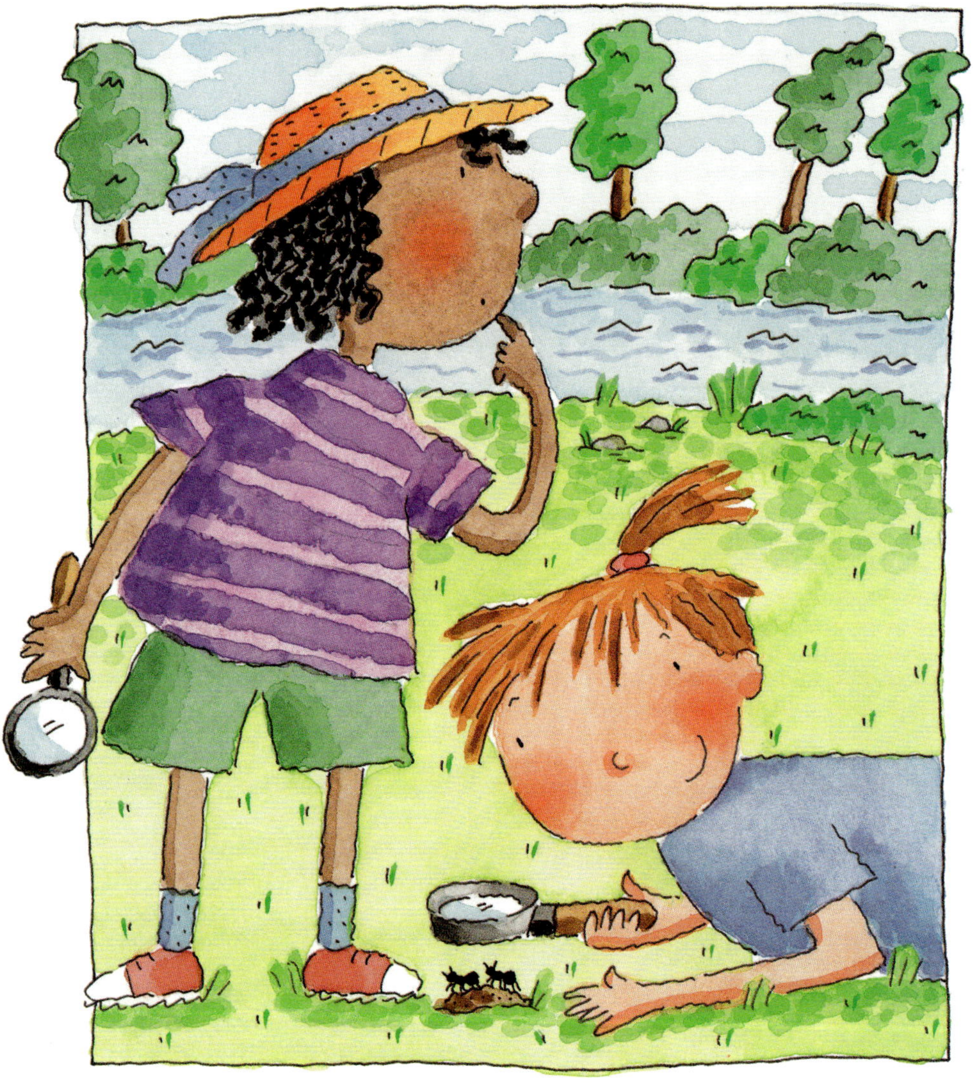

I hunt bugs.

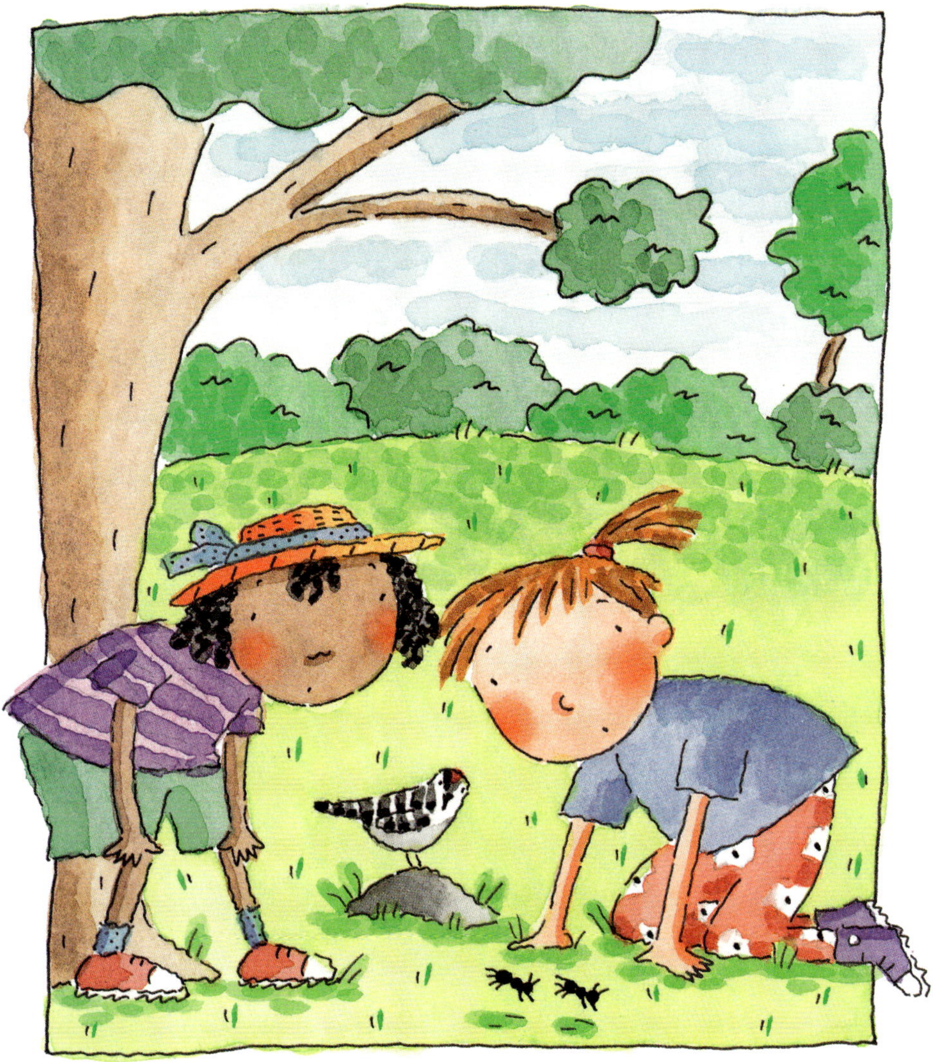

Bugs run, but not from us.

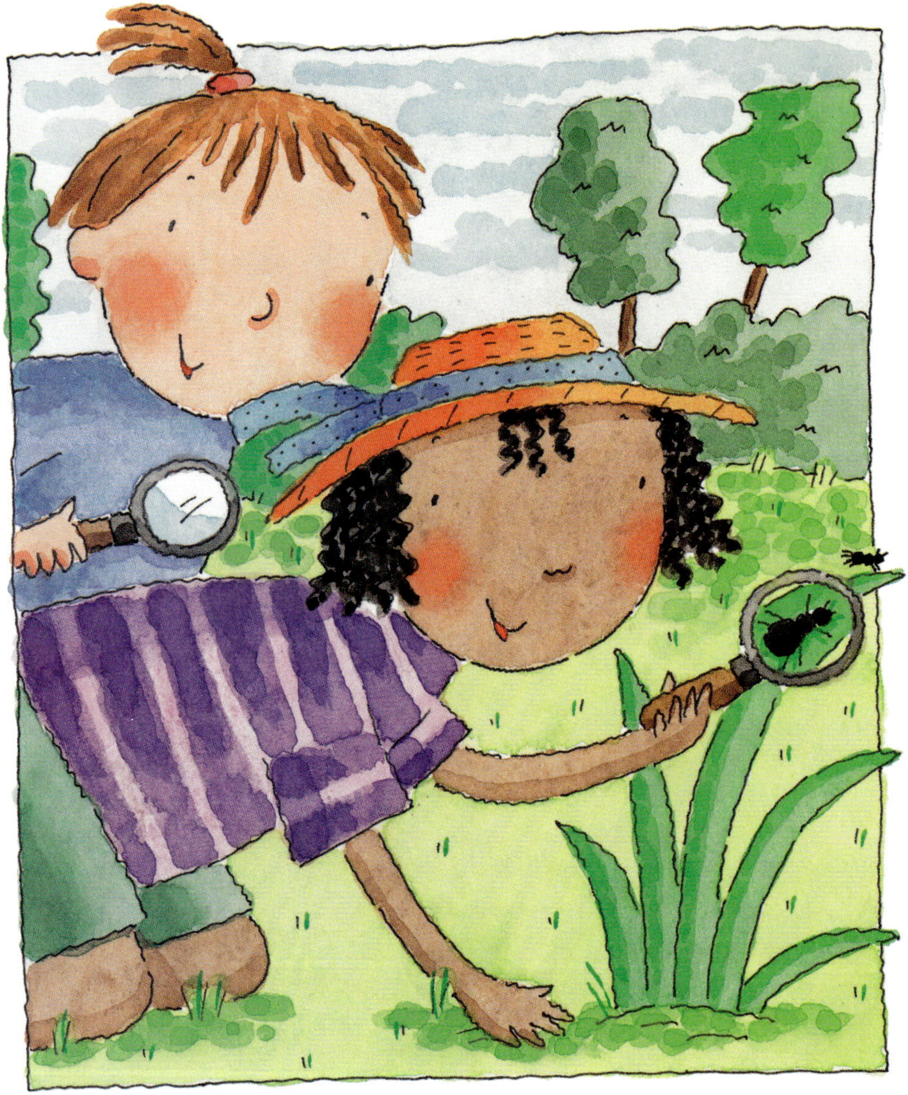

The bugs run up and up!

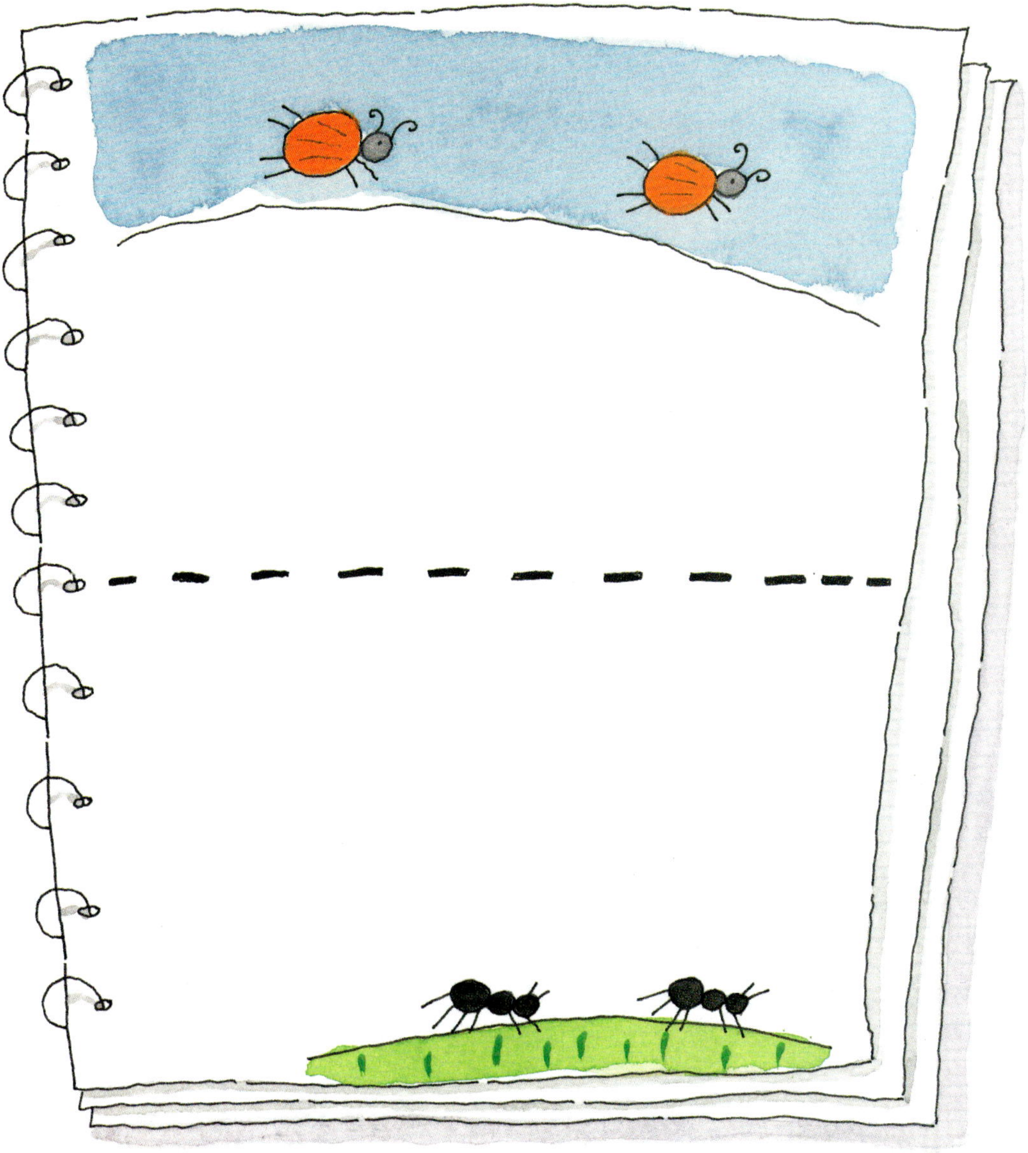

Four bugs run.

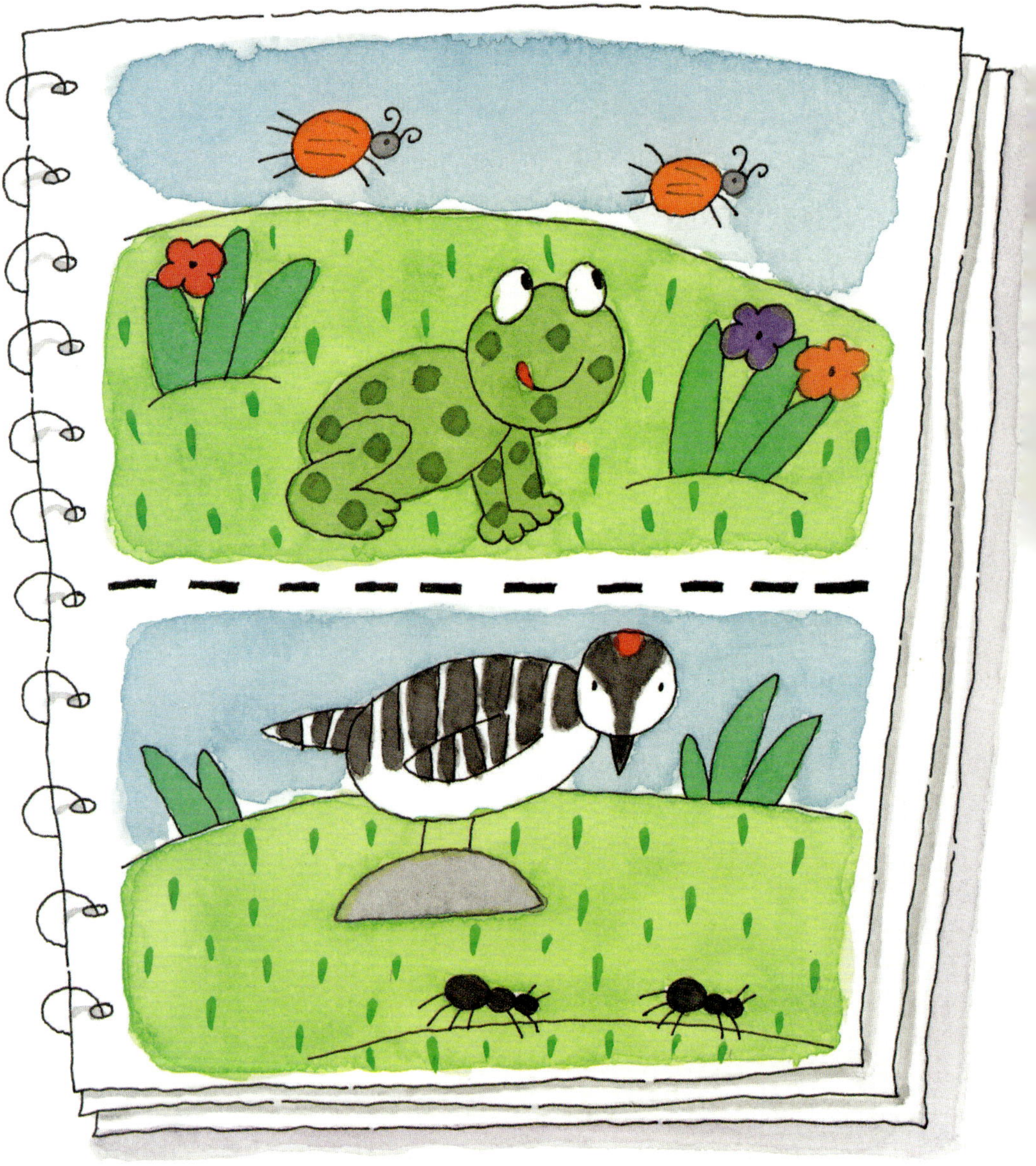

Four bugs run, but not from us.